Annals of the ICRP

Published on behalf of the International Commission on Radiological Protection

Aims and Scope

The International Commission on Radiological Protection (ICRP) is the primary body in protection against ionising radiation. ICRP is a registered charity and is thus an independent non-governmental organisation created by the 1928 International Congress of Radiology to advance for the public benefit the science of radiological protection. The ICRP provides recommendations and guidance on protection against the risks associated with ionising radiation, from artificial sources widely used in medicine, general industry and nuclear enterprises, and from naturally occurring sources. These reports and recommendations are published approximately four times each year on behalf of the ICRP as the journal *Annals of the ICRP*. Each issue provides in-depth coverage of a specific subject area.

Subscribers to the journal receive each new report as soon as it appears so that they are kept up to date on the latest developments in this important field. While many subscribers prefer to acquire a complete set of ICRP reports and recommendations, single issues of the journal are also available separately for those individuals and organizations needing a single report covering their own field of interest. Please order through your bookseller, subscription agent, or direct from the publisher.

ICRP is composed of a Main Commission, a Scientific Secretariat, and five standing Committees on: radiation effects, doses from radiation exposure, protection in medicine, the application of ICRP recommendations, and protection of the environment. The Main Commission consists of a Chair and twelve other members. Committees typically comprise 10–15 members.

ICRP uses Working Parties to develop ideas and Task Groups to prepare its reports. A Task Group is usually chaired by an ICRP Committee member and usually contains a number of specialists from outside ICRP. Thus, ICRP is an independent international network of specialists in various fields of radiological protection. At any one time, about two hundred eminent scientists and policy makers are actively involved in the work of ICRP. The Task Groups are assigned the responsibility for drafting documents on various subjects, which are reviewed and finally approved by the Main Commission. These documents are then published as the *Annals of the ICRP*.

Annals of the ICRP

ICRP PUBLICATION 122

Radiological Protection in Geological Disposal of Long-lived Solid Radioactive Waste

Editor-in-Chief
C.H. CLEMENT

Associate Editor
M. SASAKI

Authors on behalf of ICRP
W. Weiss, C-M. Larsson, C. McKenney, J-P. Minon, S. Mobbs,
T. Schneider, H. Umeki, W. Hilden, C. Pescatore, M. Vesterlind

PUBLISHED FOR

The International Commission on Radiological Protection

by

Los Angeles | London | New Delhi
Singapore | Washington DC

Please cite this issue as 'ICRP, 2013. Radiological protection in geological disposal of long-lived solid radioactive waste. ICRP Publication 122. Ann. ICRP 42(3).'

1

CONTENTS

SAGE

ICRP Publication 122

Annals of the ICRP

EDITORIAL

RADIOLOGICAL PROTECTION IN RADIOACTIVE WASTE MANAGEMENT: THE LONG VIEW

The current publication was developed as a joint effort between the radiological protection community and the waste management community. Not only has this collaboration resulted in a well-informed and useful publication, it has also brought these two communities closer together by fostering a common understanding of the basic concepts and the language used.

Disposal of radioactive waste has been the primary subject of more ICRP publications than any other phase in the nuclear fuel cycle. This likely reflects not only the importance of the subject, but also the unusual and challenging radiological protection considerations entailed.

Radioactive waste was first mentioned, if only in passing, in *Publication 7* (ICRP, 1966). The first ICRP publication dedicated specifically to radioactive waste disposal issues – *Publication 46* – came two decades later (ICRP, 1985). In this publication, two special features of radiological protection for radioactive waste disposal were identified: the probabilistic nature of future exposures, and the long time scales involved. These challenges remain an important focus in *Publication 77* (ICRP, 1997b), *Publication 81* (ICRP, 1998), and the present publication.

The long time scale involved in radioactive waste management complicates assessment of the appropriate level of protection, and questions the relationship between dose and risk in the long term. *Publication 103* (ICRP, 2007) acknowledges this difficulty, and notes that 'dose estimates should not be regarded as measures of health detriment beyond times of around several hundreds of years into the future. Rather, they represent indicators of the protection afforded by the disposal system'.

The present publication builds on the previous publications related to radioactive waste management, and applies the current system of radiological protection of *Publication 103* (ICRP, 2007) to geological disposal. Here, a key ethical principle is that of protection of future generations. Geological disposal involves time scales beyond which societal evolution can reasonably be predicted. Thus, on one hand, it is impossible to predict the evolution of social conditions and expectations, while on the other hand, the current generation has a moral obligation to provide some reasonable level of protection to future generations. Given this conundrum, what seems

5

most appropriate is to rely on 'the basic principle that individuals and populations in the future should be afforded at least the same level of protection as the current generation' (ICRP, 1998).

In addition, this publication introduces the concept of oversight or 'watchful care' during different phases of waste management and disposal. This is a crucial factor, influencing how the system of radiological protection is applied over long periods of time, and referring not only to monitoring but also to decisions on actions and implementation of plans.

The present publication is not the end of the collaboration between the radiological protection community and the waste management community, but rather an early step on the path of further discussions and cooperation. Other types of waste and management options will require different approaches for the protection of workers, the public, and the environment. As such, this report is not the last word from ICRP on the subject of waste management. Although no formal decision has been made to date, collaboration to develop a companion report to cover other radioactive wastes and management options is under consideration for the near future.

CHRISTOPHER H. CLEMENT
ICRP SCIENTIFIC SECRETARY
EDITOR-IN-CHIEF

ICRP Publication 122

Annals of the ICRP

Radiological Protection in Geological Disposal of Long-Lived Solid Radioactive Waste

ICRP PUBLICATION 122

Approved by the Commission in April 2012

Abstract–This report updates and consolidates previous recommendations of the International Commission on Radiological Protection (ICRP) related to solid waste disposal (ICRP, 1985, 1997b, 1998). The recommendations given apply specifically to geological disposal of long-lived solid radioactive waste. The report explains how the ICRP system of radiological protection described in *Publication 103* (ICRP, 2007) can be applied in the context of the geological disposal of long-lived solid radioactive waste. Although the report is written as a standalone document, previous ICRP recommendations not dealt with in depth in the report are still valid.

The 2007 ICRP system of radiological protection evolves from the previous process-based protection approach relying on the distinction between practices and interventions by moving to an approach based on the distinction between three types of exposure situation: planned, emergency and existing. The Recommendations maintains the Commission's three fundamental principles of radiological protection namely: justification, optimisation of protection and the application of dose limits. They also maintain the current individual dose limits for effective dose and equivalent dose from all regulated sources in planned exposure situations. They re-enforce the principle of optimisation of radiological protection, which applies in a similar way to all exposure situations, subject to restrictions on individual doses: constraints for planned exposure situations, and reference levels for emergency and existing exposure situations. The Recommendations also include an approach for developing a framework to demonstrate radiological protection of the environment.

This report describes the different stages in the life time of a geological disposal facility, and addresses the application of relevant radiological protection principles for each stage depending on the various exposure situations that can be encountered. In particular, the crucial factor that influences the application of the protection system over the different phases in the life time of a disposal facility is the level of oversight or 'watchful care' that is present. The level of oversight affects the capability to control the source, i.e. the waste and the repository, and to avoid or reduce potential exposures. Three main time frames are considered: time of direct oversight, when the disposal facility is being implemented and is under active supervision; time of indirect oversight, when the disposal facility is sealed and oversight is being exercised by regulators or special administrative bodies or society at large to provide additional assurance on behalf of society; and time of no oversight, when oversight is no longer exercised in case memory of the disposal facility is lost.

Keywords: Geological disposal; Radioactive waste; Protecting future generations

W. WEISS, C-M. LARSSON, C. MCKENNEY, J-P. MINON, S. MOBBS, T. SCHNEIDER, H. UMEKI, W. HILDEN, C. PESCATORE, M. VESTERLIND

PREFACE

On 21 January 2010, the Main Commission of the International Commission on Radiological Protection (ICRP) approved the creation of a new Task Group, reporting to Committee 4, to develop a report describing how the recommendations given in *Publication 103* (ICRP, 2007) can be applied in the context of the geological disposal of long-lived solid radioactive waste. This report should cover both the protection of humans (occupationally exposed workers and members of the public) and the environment, and discusses key issues such as the transition from a planned to an existing exposure situation in case of a loss of control of the waste system, as well as the applicability of dose estimated for the distant future for decision-aiding purposes. This report should also update *Publication 81* (ICRP, 1998) and provide guidance on:

- the basic concepts and terms, e.g. radiological protection principles, different types of exposure situations, dose and risk constraints, reference levels;
- the nature and role of optimisation of protection: stepwise approach, short term vs long term,
- the use and application of dosimetric units and concepts: dose and risk constraints, potential exposures, collective dose;
- the role of stakeholder involvement in different stages of planning and development.

The membership of the Task Group was as follows:

W. Weiss (Chair)   C-M. Larsson   C. McKenney
J-P. Minon   S. Mobbs   T. Schneider
H. Umeki

The corresponding members were:

W. Hilden   C. Pescatore   M. Vesterlind

Committee 4 critical reviewers were:

P. Carboneras   A. Janssens

Main Commission reviewers were:

A. Gonzalez   R. Pentreath

The Task Group wishes to thank the organisations and staff that made facilities and support available for its meetings. These include Bundesamt für Strahlenschutz (BfS), Australian Radiation Protection and Nuclear Safety Agency (ARPANSA),

9

Nuclear Regulatory Commission (NRC), Organisme national des déchets radioactifs et des matières fissiles enrichies/ Nationale Instelling voor Radioactief Aval en verijkte Splijtstoffen, (ONDRAF/NIRAS), Health Protection Agency (HPA), Le Centre d'étude sur l'Evaluation de la Protection dans le domaine Nucléaire (CEPN), Japan Atomic Energy Agency (JAEA), European Commission (EC), the Nuclear Energy Agency of the Organization for Economic Co-operation and Development (OECD/NEA), and International Atomic Energy Agency (IAEA). The Task Group is grateful to all individuals and organisations who provided valuable feedback during the web consultation.

The report was approved by the Commission in Versailles in April 2012.

MAIN POINTS

- This report provides advice on application of the Commission's 2007 Recommendations (ICRP, 2007) for the protection of humans and the environment against any
  harm that may result from the geological disposal of long-lived solid radioactive
  waste.

- For the protection of the future generations, the Commission recommendations continue to rely on the basic principle that: 'individuals and populations in the future
  should be afforded at least the same level of protection as the current generation'
  (ICRP, 1998).

- The Commission views the potential exposures to humans and the environment associated with the expected evolution of the geological disposal of long-lived solid radioactive as a planned exposure.

- Application of the protection system is influenced by the level of oversight or 'watchful care' of the disposal facility. Three main time frames have to be considered: time
  of direct oversight, when the disposal facility is being operated and is under active
  supervision; time of indirect oversight, when the disposal facility is partly or fully
  sealed where indirect regulatory, administrative or societal oversight might continue;
  and time of no oversight, when the memory of the disposal facility has been lost.

- If oversight ceases to exist in the post closure period, the disposal system is still a
  functioning facility and potential exposures should be considered as planned.

- The different decisions to be made relating to the evolution of oversight should be
  discussed with stakeholders.

- For application of the justification principle, waste management and disposal operations have to be considered as an integral part of the practice generating the waste.
  This justification should be reviewed over the life time of that practice whenever new
  and important information becomes available.

- As stated in previous publications on radioactive waste management (ICRP, 1997b,
  1998), the control of public exposure in the distant future through a process of constrained optimisation will obviate the direct use of individual dose limits.

- Optimisation of protection is the central element of the stepwise design, construction, and operation of a geological disposal facility.

- Optimisation has to be understood in the broadest sense as an iterative, systematic,
  and transparent evaluation of protective option, including Best Available
  Techniques, for enhancing the protective capabilities of the system and reducing
  its potential impacts (radiological and others).

- In application of the optimisation principle, the radiological criterion for the design of a waste disposal facility recommended by ICRP is an annual dose constraint for the population of 0.3 mSv year^{-1} and below the annual dose limit of 20 mSv year^{-1} or 100 mSv in 5 years for occupationally exposed workers.

- A risk constraint for the population of 1×10^{-5} year^{-1} is recommended when applying an aggregated approach combining probability of the exposure scenario and the associated dose.

- In the very long term, dose and risk criteria should be used for the comparison of options rather than a means of assessing health detriment.

- For natural events included in the design-basis evolution, the Commission recommends selection of dose or risk constraints in the band for planned exposure situations.

- For severe natural disruptive events not taken into account in the design-basis evolution and inadvertent human intrusion, application of the risk or dose constraint does not apply. In that case, if the events were to occur while there is still (direct or indirect) oversight of the disposal facility, the ensuing exposure situation (emergency or existing) should be considered by the competent authority, and the relevant protection measures should be implemented.

- For inadvertent human intrusion, the design and siting of the facility should include features to reduce the possibility of such events.

- Judgement of the quality of the system design developed or implemented has to be made, and reviewed critically when needed, in a well-structured and transparent process, with the involvement of all relevant stakeholders.

- General implementation of the Commission's recommendations requires a management system that integrates safety, health, environmental, security, quality, and economic elements, with safety being the fundamental goal.

- For planned exposure situations, doses should be assessed on the basis of the annual dose to the representative person.

- Consideration of environmental protection, where appropriate, should be part of the risk-informed decision making.

EXECUTIVE SUMMARY

(a) This report provides advice on application of the Commission's 2007 Recommendations (ICRP, 2007) for the protection of humans and the environment against any harm that may result from the geological disposal of long-lived solid radioactive waste. It illustrates how the key protection concepts and principles of *Publication 103* (ICRP, 2007) should be interpreted, and how they apply over the different time frames a geological disposal facility for long-lived solid radioactive waste would have to provide radiological protection (see Fig. A).

(b) The goal of a geological disposal facility is to contain and isolate the waste in order to protect humans and the environment for time scales that are comparable with geological time scales. At large distances from the surface, changes are particularly slow. Given the distance from the surface and the selection of appropriate sites, the potential for human intrusion is limited. Radioactivity is increasing with time, and any release will be delayed and further diluted by a properly chosen geological formation. Geological disposal is recognised by international organisations as especially suited for high-level radioactive waste or spent fuel where long-term containment and isolation is required. Geological disposal may also be used for other long-lived wastes, especially when a similar need for long-term protection applies.

(c) One of the important factors that influence application of the protection system over the different phases of the life time of a geological disposal facility is the level of oversight or 'watchful care' of the disposal facility. The level of oversight directly affects the capability to control the source, and to avoid or reduce some exposures. Three main time frames have to be considered: time of direct oversight, when the disposal facility is being operated and active control is taking place (operational phase); time of indirect oversight, when the disposal facility is partly (backfilling and sealing of drifts) or fully sealed (postclosure period) where indirect regulatory or societal oversight might continue for a period and then be supplemented or replaced by indirect oversight (e.g. monitoring of the performances of the repository and the pathways for potential radionuclide releases, verification that restrictions on land control use are being met, maintaining records and memory of the facility, etc.);

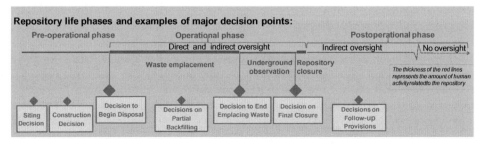

Fig. A. Disposal facility phases and relevant oversight periods.

and time of no oversight (postclosure period), when the memory of the disposal facility is lost. In the periods of indirect or no oversight, once the facility is sealed, protection relies on the passive controls built into the facility at the time of its design, licensing, and operation.

(d) The design and associated safety case of a geological disposal facility address a series of evolutions with different probabilities that may be defined by regulation. Besides these design-basis evolutions, the developer/implementer, overseen by the regulator and society, may want to assess evolutions in non-design-basis conditions in order to judge the robustness of the facility.

(e) This report describes the radiological concepts and criteria that should be used by the designer and/or operator of the facility, the regulator, and the concerned stakeholders. Various dose and risk constraints are used for assessment of the safety and radiological protection of a geological disposal facility for long-lived radioactive waste. Optimisation addresses the main aim of a disposal facility (i.e. the radiological protection of humans and the environment). Optimisation of protection is the central element of the stepwise construction and implementation of a geological disposal facility. It has to cover all elements of the system, including the societal component, in an integrated way. Important aspects of optimisation of protection must occur prior to waste emplacement, largely during the siting and design phase. The optimisation efforts can be informed by, and construction supplemented with, consideration of Best Available Techniques (BAT) as applied to all stages of disposal facility siting and design. During the implementation phase, some further optimisation is possible, but it is accepted that very little can be done to further optimise the performance of the engineered features after waste emplacement has occurred, and more so when galleries have been sealed.

(f) In the distant future, the geological disposal facility might give rise to some releases to the accessible environment, and the safety case has to demonstrate that such releases, should they occur, will be within radiological protection criteria specified as part of the regulatory requirements. In application of the optimisation principle, the reference radiological impact criterion for the design of a waste disposal facility recommended by ICRP is an annual dose constraint for the population of 0.3 mSv year^{-1} (ICRP, 2007), without any weighting of doses in the distant future. For doses in the future and for less likely events resulting in exposures, both categorised as potential exposures, the Commission continues to recommend a risk[1] constraint for the population of 1×10^{-5} year^{-1} when applying an aggregated approach combining probability of the exposure scenario and the associated dose. However, *Publication 103* (ICRP, 2007) also warns that effective dose loses its direct connection to health detriment for doses in the future after a time span of a few generations, given the evolution of society, human habits, and characteristics. Furthermore, in the distant future, the geosphere, the engineered system and, even more so, the biosphere will evolve in a less predictable way. The scientific basis for assessments of detriment to health at very long times into the future therefore becomes

---

[1] Risk is used in this document to mean 'radiological risk' as defined in *Publication 103* (ICRP, 2007).

uncertain, and the strict application of numerical criteria may be inappropriate. In the very long term, dose and risk criteria should be used for the comparison of options rather than a means of assessing health detriment.

(g) The design-basis evolution of the geological disposal facility includes the expected evolution of the protection provided by the facility, and also events with a low probability of occurrence (less likely evolutions). It does not include either severe disturbing events of very low probability that may disrupt the facility, or inadvertent human intrusion. The exposures arising from the design-basis evolution scenarios are planned exposure situations as defined in *Publication 103* (ICRP, 2007). They include potential exposures from events with low probability, which have to be considered as part of the design basis. More specifically, for exposure to be delivered in the distant future, potential exposures will have to be considered due to the considerable uncertainties surrounding such exposures (ICRP, 2007, Para. 265). If severe disturbing events outside the design basis occur while there is still oversight (direct or indirect) of the disposal facility, the ensuing situation will be considered by the competent authority at that time, and the relevant protection measures will be implemented. If a severe disturbing event occurs when there is no longer oversight of the disposal facility, there is no certainty that a competent authority will be able to understand the source of the exposure, and therefore it is not possible to consider with certainty the implementation of relevant measures to control the source. Inadvertent human intrusion into the geological disposal facility is not a relevant scenario during the period of direct or indirect oversight. In the period of no oversight, inadvertent human intrusion may occur and the consequences considered by the competent authorities at that time, if and when they understand the source of the exposure.

(h) For the design-basis evolution, the dosimetric criteria relevant to planned exposure situations are considered for assessing the safety and robustness of the disposal facility over the three main time frames. In the design stage, potential impacts of severe disturbing events may be estimated using stylised or simplified calculations. An indication of the robustness of the system could then be obtained by comparing these results with numerical values of dose or risk, if required. If this approach is adopted, the appropriate reference levels should be those for an existing exposure situation (a few mSv per year), or for an emergency exposure (in the range of 20–100 mSv for the first year), depending on the specific scenario. If the event actually occurs in the future, the competent authority should apply the relevant protection criteria at the time.

(i) The safety case of a geological disposal facility, by including events of low probability and exposures to be delivered in the distant future, includes consideration of how to deal with potential exposures as defined by *Publication 103* (ICRP, 2007).

(j) ICRP recommends that dose or risk estimates derived from these exposure assessments should not be regarded as direct measures of health effects beyond time scales of around several hundred years into the future. Rather, they represent indicators of the protection afforded by the geological disposal facility.

(k) Application of the three exposure situations and of dose limits, constraints, and reference levels as defined in *Publication 103* (ICRP, 2007) during the three main time frames is indicated in Table A.

Table A. Radiological exposure situations as function of disposal facility evolution, and presence and type of oversight.

Disposal facility status	Type of oversight		
	Direct oversight	Indirect oversight	No oversight
Design-basis[*] evolution	Planned (normal and potential) exposure situation[†]	Planned (potential) exposure situation[†,‡]	Planned (potential) exposure situation[†,‡]
Non-design-basis evolution[§]	Emergency exposure situation at the time of exposure, followed by an existing exposure situation	Emergency exposure situation at the time of exposure, followed by an existing exposure situation[¶,**]	Emergency and/or existing exposure situation, once exposure is recognised[¶,**]
Inadvertent human intrusion	Not relevant	Not relevant	Emergency and/or existing exposure situation, once exposure is recognised[¶,**]

[*] The design basis is the envelope of both normal and potential exposures that are used in designing the facility.

[†] In the planning phase: 20 mSv year^{-1} dose limit to workers and dose constraint as specified by the operator; 1 mSv year^{-1} dose limit for public exposures from all sources and 0.3 mSv year^{-1} dose constraint for waste disposal. For potential exposure of the public in case of the application of an aggregated approach, a risk constraint of 1×10^{-5} year^{-1} is recommended.

[‡] No worker dose is foreseen during the period of indirect or no oversight. Releases in the distant future give rise to potential exposure (ICRP, 2007, Para. 265). Comparisons with the dose or risk constraint become increasingly less useful for compliance purposes at times further in the future.

[§] Non-design-basis evolutions include very unlikely or extreme events that could result in significant exposure to humans and the environment.

[¶] If such an event were to occur in the future, the competent authorities of the time would assess whether it had resulted in an emergency exposure situation or in an existing exposure situation, or the equivalent categories of exposure at that time. If *Publication 103* (ICRP, 2007) was still extant, it would be expected that the reference levels for emergency and/or existing exposure situations would be applied, as appropriate. In the period of no oversight, the exposure may not be recognised immediately.

[**] At the planning stage, the potential radiological impact is typically evaluated using stylised or simplified scenarios. The results of those analyses can be used as indicators of system robustness by comparing them with numerical values. In that case, application of the reference levels defined for emergency and/or existing exposure situations is recommended. It should be noted that a fully optimised system may result in a distribution of doses where some are above the reference level (ICRP, 2009a, p. 37).

GLOSSARY

Best Available Techniques (BAT)

'Best available techniques' mean the most effective and advanced stage in the development of activities and their methods of operation designed to prevent and, where not practicable, to reduce emissions and the impact on the environment as a whole.

Committed effective dose, $E(\tau)$

The sum of the products of the committed organ- or tissue-equivalent doses and the appropriate tissue weighting factors (w_T), where τ is the integration time in years following the intake. The commitment period is taken to be 50 years for adults, and to 70 years of age for children.

Committed equivalent dose, $H_T(\tau)$

The time integral of the equivalent dose rate in a particular tissue or organ that will be received by an individual following intake of radioactive material into the body by a Reference Person, where τ is the integration time in years.

Containment

The function of confining the radionuclides within the manmade barriers that either constitute the waste form or that separate the waste form from the host geological formation.

Detriment

The total harm to health experienced by an exposed group and its descendants as a result of the group's exposure to a radiation source. Detriment is a multidimensional concept. Its principal components are the stochastic quantities: probability of attributable fatal cancer, weighted probability of attributable non-fatal cancer, weighted probability of severe heritable effects, and length of life lost if the harm occurs.

Direct oversight, see Oversight

Disposal facility

An engineered facility for the disposal of spent fuel or radioactive waste, without the intention of retrieval. It includes the entire underground construction (tunnels, caverns, and access shafts), the emplaced waste, and the sealing and backfill materials.

Dose

The sum of the committed effective dose from intakes and the effective dose from external irradiation.

Dose constraint

A prospective and source-related restriction on the individual dose from a source that provides a basic level of protection for the most highly exposed individuals from a source, and serves as an upper bound on the dose in optimisation of protection for that source. For occupational exposure, the dose constraint is a value of individual dose used to limit the range of options considered in the process of optimisation. For public exposure, the dose constraint is an upper bound on the annual doses that members of the public should receive from the planned operation of any controlled source.

Dose limit

The value of the effective dose or the equivalent dose to individuals from planned exposure situations that shall not be exceeded. Dose of record is assigned to the worker for purposes of recording, reporting, and retrospective demonstration of compliance with regulatory dose limits.

Effective dose

The tissue-weighted sum of the equivalent doses in all specified tissues and organs of the body, given by the expression:

$$E = \sum_T w_T \sum_R w_R D_{T,R} \quad \text{or} \quad E = \sum_T w_T H_T$$

where H_T or $w_R D_{T,R}$ is the equivalent dose in a tissue or organ T, and w_T is the tissue weighting factor. The unit for effective dose is the same as for absorbed dose, J kg^{-1}, and its special name is sievert (Sv). The main and primary uses of effective dose in radiological protection for both occupational workers and the general public are (ICRP, 2007, Para. 153): prospective dose assessment for planning and optimisation of protection; and retrospective dose assessment for demonstrating compliance with dose limits, or for comparing with dose constraints or reference levels. In practical radiological protection applications, effective dose is used for managing the risks of stochastic effects in workers and the public.

Emergency exposure situation

Emergency exposure situations are exposure situations resulting from a loss of control of a planned source, or from any unexpected situation (e.g. a malevolent event) which require urgent action to avoid or reduce undesirable exposures.

Equivalent dose, H_T

The dose in a tissue or organ T given by:

$$H_T = \sum_R w_R D_{T,R}$$

where $D_{T,R}$ is the mean absorbed dose from radiation R in a tissue or organ T, and w_R is the radiation weighting factor. Since w_R is dimensionless, the unit for the equivalent dose is the same as for absorbed dose, J kg^{-1}, and its special name is sievert (Sv).

Existing exposure situation

Existing exposure situations are exposure situations resulting from sources that already exist when a decision to control them is taken (natural radiation, past activities or after emergencies).

Exposure situation

An exposure situation is the process that includes a natural or man-made radiation source, the transfer of the radiation through various pathways and the exposure of individuals.

Hazard

A property or situation that in certain circumstances could lead to harm. Hazard is the potential to cause harm and is distinct from risk (see below), which defines the likelihood of harm occurring from a defined set of circumstances.

Indirect oversight, see Oversight

Isolation

The function of preventing the release of radionuclides to the living environment in quantities that exceed predefined constraints.

Justification

The process of determining whether either: (1) a planned activity involving radiation is, overall, beneficial [i.e. whether the benefits to individuals and to society from introducing or continuing the activity outweigh the harm (including radiation detriment) resulting from the activity]; or (2) the decision to control exposure in an emergency or an existing exposure situation is likely, overall, to be beneficial [i.e. whether the benefits to individuals and to society (including the reduction in radiation detriment) outweigh its cost and any harm or damage it causes].

Occupationally exposed worker

Any person who is employed, whether full time, part time, or temporarily, by an employer, and who has recognised rights and duties in relation to occupational radiological protection.

Optimisation of protection

The process to keep the likelihood of incurring exposures, the number of people exposed, and the magnitude of their individual doses as low as reasonably achievable, taking into account economic and societal factors.

Oversight

Oversight is a general term for 'watchful care' and refers to society 'keeping an eye' on the technical system and the actual implementation of plans and decisions. It includes regulatory supervision, in the form of control and inspection, preservation of societal records, and societal memory of the presence of the facility. Three time periods are considered for oversight.

- Direct oversight refers to active control measures during the operational phase of the facility e.g. inspections and monitoring.
- Indirect oversight refers to measures that are used once the facility is closed and there is no longer access to the underground facilities e.g. a period of continued regulatory control, preservation of land use records, monitoring by society to check that the environmental conditions are not degrading.
- No oversight refers to situations when the memory of the presence of the disposal facility is lost and society no longer keep a watchful eye on the facility.

Planned exposure situations

Planned exposure situations are exposure situations resulting from the operation of deliberately introduced sources.

Potential exposure

Exposure that is not expected to be delivered with certainty but that may result from an accident at a source or an event or sequence of events of a probabilistic nature, including equipment failures and operating errors. Due to the large uncertainties surrounding exposures that may occur in the future, they are considered as potential exposures.

Reference animals and plants

Hypothetical entities with assumed basic biological characteristics of a particular type of animal or plant, as described to the generality of the taxonomic level of Family, with defined anatomical, physiological, and life-history properties.

Reference level

In emergency or existing controllable exposure situations, this represents the level of dose or risk above which it is judged to be inappropriate to plan to allow exposures to occur, and below which optimisation of protection should be implemented. The chosen value for a reference level will depend upon the prevailing circumstances of the exposure under consideration. A reference level is not a limit or constraint, and an optimised system may result in a distribution of doses, with some doses above the reference level.

Representative person

An individual receiving a dose that is representative of the more highly exposed individuals in the population (ICRP, 2006). This term is the equivalent of, and replaces, 'average member of the critical group' described in previous ICRP recommendations.

Retrievability

The ability, in principle, to recover waste or entire waste packages once they have been emplaced in a repository; retrieval is the concrete action of removal of the waste. Retrievability implies making provisions in order to allow retrieval should it be required.

Reversibility

The ability, in principle, to reverse or reconsider decisions taken during the progressive implementation of a disposal system; reversal is the concrete action of overturning a decision and moving back to a previous situation.

Risk

The probability of harmful or injurious consequences (e.g. cancer) associated with exposures or potential exposures in a year. It takes into account the probability of receiving a dose in a year, and the probability that the dose received will give rise to harm. Risk = likelihood of occurrence × seriousness if incident occurs.

Safety case

A safety case is a structured set of arguments and evidence demonstrating the safety of a system. More specifically, a safety case aims to show that specific targets and criteria are met.

Stakeholder(s)

Parties who have interest in and concern about a given situation. Examples include the exposed individuals (either workers or members of the public) or their representatives (trade unions, local associations), institutional and non-institutional technical support to the decision-making process (approved dosimetric services, qualified experts, formal technical services, public expert organisations, private laboratories), and representatives of the society, either by an elective process (elected representatives) or a participative process (environmental associations).

Storage

The holding of spent fuel or radioactive waste in a facility that provides for its containment, with the intention of retrieval.

1. INTRODUCTION

(1) This report is written as a standalone presentation of the 2007 ICRP system of radiological protection (ICRP, 2007) as it should be applied in the context of geological disposal of long-lived solid radioactive waste. It covers all issues related to radiological protection of humans and the environment from harm following the geological disposal of long-lived solid radioactive waste. Nevertheless, the previous ICRP recommendations not dealt with in depth in the report are still valid. Where the Commission's recommendations are unchanged, or issues are addressed sufficiently in publications by other international organisations, references are given and no detailed discussion is provided. Although this report deals specifically with geological disposal, many of the recommendations may influence decision making during storage of radioactive waste pending disposal solutions, and are also, in many respects, relevant to near-surface disposal of radioactive waste.

(2) The occupational exposure of workers and exposure of the public are managed in accordance with the ICRP system of radiological protection. The main protection issue dealt with in this report concerns exposures that may or may not occur in the distant future. Any corresponding estimates of doses and risks to individuals and populations will have growing associated uncertainties as a function of time due to incomplete knowledge of future disposal facility behaviour, geological and biospheric conditions, social and economic conditions, and human habits and characteristics. Furthermore, due to the long time scales, verification that protection is being achieved cannot be envisaged in the same manner as for current discharges because knowledge of the presence of the disposal facility may eventually be lost. Neither can it be assumed that effective mitigation measures will necessarily be carried out, should they be required in the distant future. Nevertheless, the ICRP system of radiological protection can be applied to the geological disposal of long-lived solid radioactive waste with due interpretation.

(3) In the context of the Commission's recommendations, waste is any material for which no further use is foreseen. Waste, as generated, includes liquid and gaseous effluents as well as solid materials. Storage means the holding of waste in a storage facility with the express intention of retrieval at a later time. Disposal means the emplacement of waste in a repository without the intention of retrieval. Waste management means the whole sequence of operations starting with the generation of waste and ending with disposal.

(4) Geological disposal is intended to contain and isolate long-lived solid radioactive waste. Containment is the function of confining the radionuclides within the manmade barriers that either constitute the waste form or that separate the waste form from the host geological formation, or else within an appropriately defined volume surrounding the repository. Isolation is the function of preventing the release of radionuclides to the living environment in quantities that exceed predefined

constraints. Geological disposal is especially suited for high-level waste, spent fuel, and intermediate-level waste containing radionuclides with long half-lives. These are wastes with high specific activities. If these wastes were not disposed of in a geological disposal facility but instead remained on the surface of the earth, special precautions would be necessary to maintain safety. This report does not address near-surface disposal facilities because they differ from geological disposal facilities in two key aspects: the waste for which they are intended, and the means by which the containment and isolation functions are achieved.

(5) Technical solutions for the permanent isolation of long-lived solid radioactive waste hundreds of metres below the surface in geological formations are being developed and pursued in a number of countries. Geological disposal is currently recognised by international organisations in charge of radioactive waste management as especially suited for high-level radioactive waste or spent fuel where long-term containment is required. Geological disposal may also be used for other wastes containing long-lived radionuclides because similar long-term protection requirements can be formulated. An example of geological disposal is the emplacement of waste in excavated tunnels or shafts, followed by backfilling and sealing the entire facility.

(6) The goal of a geological disposal facility is to achieve containment and isolation of the waste, and to protect humans and the environment for extended time scales over which climatic and surface changes may occur. At great distances from the surface, geological formations can be identified that exhibit very long stable geological conditions (i.e. many thousands to millions of years). A properly chosen geological formation assures stable chemical and physical conditions for the waste, and will reduce and delay any releases of radionuclides to the geosphere. In this context, 'distance' can imply horizontal or vertical distance as, for example, the case of a disposal facility sited deep within a mountain. Additionally, if a site is chosen in an area with no known natural resources, the likelihood for inadvertent human intrusion into the facility may be limited.

(7) The safety strategy implemented for geological disposal is to contain and isolate the waste. No exposure is ever intended, although very low levels of exposure may happen in the distant future. The disposal facility is thus to be seen as a functional facility whose safety and protection are in-built and continue after facility closure. This allows radioactive decay to take place, and delay the eventual release of any contaminants to the biosphere and the environment. Furthermore, isolation reduces the risk of inadvertent human intrusion into the waste.

(8) Geological disposal of long-lived solid radioactive waste poses a number of challenges related to radiological protection, such as the nature and role of optimisation of protection during the various phases of the development and implementation of the disposal facility, and the applicability of dose and risk indicators in the distant future for decision aiding. This report explains how the fundamental radiological protection principles as laid out in *Publication 103* (ICRP, 2007) can be

applied under these circumstances. It also considers elements that can assist in demonstrating compliance with these principles, and how in broad terms, they relate to other protective goals that are considered in environmental impact assessments.

2. SCOPE

(9) The Commission has previously published recommendations for the disposal of long-lived radioactive waste (ICRP, 1985, 1997b, 1998) consistent with the general recommendations for the application of its previous overall system of radiological protection (ICRP, 1991). More recently, the Commission has published new general recommendations (ICRP, 2007). This report summarises and explains how the 2007 Recommendations specifically apply to a geological disposal facility for long-lived solid radioactive waste.

(10) This report deals with the radiological protection of humans and the environment following the disposal of long-lived solid radioactive waste in geological disposal facilities. The recommendations given in this report apply to disposal facilities where there is still an opportunity for their implementation during the site selection, design, construction, and operational phases. They should also be taken into account in the justification of practices generating waste. Optimisation on the basis of radiological criteria is an important part of the overall optimisation process during the design and operation of the disposal facility at specific periods and for specific aspects of the disposal facility.

(11) The report does not describe the disposal safety assessment in detail. It rather provides a description of how protection criteria can be used in the safety analysis, and establishes recommendations on protection issues related to the disposal of long-lived solid radioactive waste.

(12) Although many considerations in this report are also relevant to near-surface disposal, it does not supersede previous ICRP recommendations (ICRP, 1998) for the radiological protection of workers, members of the public, and the environment in the case of near-surface disposal facilities or other disposal options, which remain valid.

(13) Radiological protection is only one set of the protection concepts that is used by safety analysts in documenting the protection capability of the disposal facility. Other concepts may relate to the protection of resources in a sustainable way such as, for example, groundwater as a drinking water resource. The chemical toxicity of the waste or the waste containment system in a disposal facility for radioactive waste may also be considered. These broader aspects are not addressed in this document.

3. BASIC VALUES, PRINCIPLES AND STRATEGIES FOR PROTECTING FUTURE GENERATIONS

3.1. Values for protecting future generations

(14) The composition of radionuclides in long-lived radioactive waste evolves over time, changing the degree of the hazard. Due to the long half-life of some radionuclides, and the ingrowth of others, some wastes continue to be a hazard into the distant future but overall the activity of waste decreases with time.

(15) Over the last decades, reflections on safety and societal issues associated with this long-term dimension of the hazard from long-lived radioactive waste clearly point out the complexity of the situation: on one hand, it is not possible to envisage how society will be organised in the distant future, while on the other hand, the current generation has a duty of care to future generations. This complexity is central to the ethical reflections required in the design of waste management strategies that are based on the precautionary principle and sustainable development in order to preserve the health and environment of future generations.

(16) In ICRP *Publication 81* the Commission recommends that 'individuals and populations in the future should be afforded at least the same level of protection as the current generation' (ICRP, 1998, Para. 40). This basic principle is broadly consistent with the requirement of the 1997 Joint Convention on the Safety of Spent Fuel Management and on the Safety of Radioactive Waste Management stating that individuals, society and the environment should be protected from the harmful effects of ionising radiation, now and in the future, in such a way that: 'the needs and aspirations of the present generation are met without compromising the ability of future generations to meet their needs and aspirations' (IAEA, 1997).

(17) In the same vein, the obligations of the present generation towards the future generation are complex, involving, for instance, not only issues of safety and protection but also transfer of knowledge and resources. Due to the technical and scientific uncertainties, and the evolution of society in the long term, it is generally acknowledged that the present generation is not able to ensure that societal action will be taken in the future, but needs to provide the means for future generations to cope with these issues.

3.2. Principles of radiological protection

(18) The ICRP protection system, as described in the 2007 Recommendations (ICRP, 2007), continues to rely on three fundamental principles: justification, optimisation of protection and application of dose limit. These applied to the three types of exposure situation considered by the Commission to organize radiological protection: planned exposure situations, emergency exposure situations and existing exposure situations.

(19) The optimisation principle is of primary importance and its role has been reinforced in the 2007 Recommendations (ICRP, 2007). For this purpose, ICRP

recommends that in assessing the level of protection for humans: 'the likelihood of incurring exposures, the number of people exposed, and the magnitude of their individual doses should all be kept as low as reasonably achievable, taking into account economic and societal factors' (ICRP, 2007, Para. 203).

(20) For these assessments, two concepts are considered by ICRP: dose and risk. Associated with dose and risk, the concept of health detriment, as introduced by ICRP in *Publication 26* (ICRP, 1977), is also a key concept to consider when assessing the level of protection. The application of this concept aims to provide an estimate of the total harm to health to individuals and their descendants as a result of an exposure, assuming a linear-non-threshold dose–effect relationship. For exposures that may occur in the distant future, the relevance and meaning of dose and risk is of interest, and their interpretation over the different phases (as defined in Section 3.3.1) has to be clarified. It should be noted that knowledge of the relationship between dose and effect may very well change in the future, as has already been demonstrated by past re-assessments of nominal probability coefficients. Likewise, the ability to cure or mitigate induced health effects may change in the future. It is not possible to make any prediction of the direction of these changes. Thus, ICRP considers that the efforts to avoid and/or reduce any effect on human health and on the environment in the distant future have to be guided entirely by the current understanding of health and environmental effects.

(21) Notwithstanding the uncertainties described above, ICRP dosimetric quantities and health detriment can be used for long-term assessment. In fact, assessment of the robustness of the protection system provided by solid waste disposal in the long term does not need a precise knowledge of the evolution of the general health status of the population in the distant future. At the design stage, focusing on the evaluation of health effects in a particular hypothetical population in the distant future is not appropriate. The challenge is rather to estimate, in an optimisation process through a comparison (using inter-alia dose and risk indicators) of alternative options, the levels of protection achieved by a given disposal facility and to judge if the estimated protection level of the chosen strategy is acceptable in the light of the level of protection accepted today.

3.3. Strategies for the management of long-lived solid radioactive waste

(22) Due to the nature and longevity of hazards, the fundamental strategy adopted for the management of long-lived radioactive waste in order to achieve the safety objective is to contain and isolate the waste from the environment for as long as possible. The goal of a geological disposal facility is to provide protection of humans and the environment from the hazards that the radioactive waste poses over time. The current generation has to consider the preservation of the resources and environment of future generations when designing the waste management strategy. This includes possible developments over different time scales with different levels of presence of human institutional control, but also the uncertainty concerning the level of presence of humans themselves.

(23) It is internationally recognised that only materials that have been declared to have no further use for society (waste) are disposed of, so that there is no intention to retrieve it, even if technical options to do so were available. Disposal is not to be confused with a storage situation. Currently, the reference option is to dispose of these wastes in engineered disposal facilities located in suitable geological formations (IAEA, 1997; OECD/NEA, 2008).

(24) A stepwise process, involving various stakeholders, may be adopted in planning for the development and implementation of a disposal facility, including final closure. In that context, provisions may be made for the operation of the facility to be reversible, and for the emplaced wastes to be retrievable, to an extent that may vary between different programmes. Retrievability does not imply an intention to retrieve, nor is retrieval a contingency plan for the disposal facility. The decision to perform any retrieval would be a separate decision taken in the future, according to the radiological principles that apply to a new planned activity. If provisions are made for reversibility or retrievability, these should not have unacceptable consequences for radiological protection. For example, there might be a proposal to keep a facility open that would otherwise be ready for closure, solely to maintain the option to retrieve waste emplaced in the facility. Safety cases would then need to demonstrate that processes such as degradation of waste packages and other unexpected events would not unacceptably affect the protection of humans or the environment.

(25) The 'contain and concentrate' strategy makes it possible, in principle, for the waste to be re-accessed either voluntarily or involuntarily at some time in the future. Therefore, disposal facilities should be designed to reduce the possibility of inadvertent human intrusion. There are, to some extent, conflicting requirements involved and a balance has to be found in each case, taking into consideration the time scales, the nature of the waste, the nature of the host geological formation, and the evolving desires of society.

3.3.1. Phases of a disposal facility and the safety analysis process

(26) The development of a geological disposal facility involves three main phases (Fig. 3.1), the durations of which vary between national programmes depending on the design and on each country's approach to decision making. These phases are associated with different types of oversight ('watchful care') of the disposal facility. Oversight refers to regulators or society 'keeping an eye' on the technical system and the actual implementation of plans and requirements. It includes regulatory supervision and control, preservation of societal records, and societal memory of the presence of the facility. Direct oversight refers to active control measures during the operational phase of the facility (e.g. inspections and monitoring). Indirect oversight refers to measures that are used once the facility is closed and there is no longer access to the underground facilities (e.g. a period of continued regulatory control, preservation of land-use records, monitoring by society to check that the environmental conditions are not degrading). Eventually, there may be a time when the memory of the presence of the disposal facility is lost and society no longer keep a watchful eye

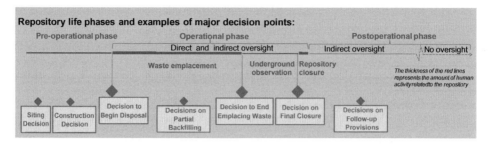

Fig. 3.1. Disposal facility phases and relevant oversight periods.

on the facility, i.e. in other words, a time when there is no oversight of the disposal facility.

(27) During the operational phase, it is expected that direct oversight of the facility is performed consistent with the oversight performed at other nuclear facilities that handle similar radioactive materials (e.g. regulatory inspections). Following closure of the facility (postoperational phase), it is expected that indirect oversight includes the monitoring of the performances of the repository and potential pathways for exposure, the preservation of records of the presence of the facility and the verification of land-use restrictions. However, the continuation of indirect oversight during the postoperational phase becomes more uncertain at later times (e.g. hundreds of years), and it may be assumed that at some point in time, memory could be lost and there is no further indirect oversight. This is one reason why geological disposal facilities are developed and designed not to rely on oversight in the distant future (i.e. passive safety), although the aim is not to lose the memory of the site.

The pre-operational phase

(28) During this phase, the disposal facility is designed, the site is selected and characterised, the manmade materials are tested and the engineering feasibility is demonstrated, safety assessments for operational and postoperational phases are developed, the licences for building and operation are applied for and received, and construction begins. A baseline monitoring programme of environmental conditions is also established.

The operational phase

(29) During this phase, the waste is emplaced, followed by a period of observation prior to closure. For a period during this phase, some galleries may be filled and sealed having reached their final configuration, while others may still be open and being filled.

(30) This phase is under direct oversight of the regulatory safety authorities in cooperation with other relevant stakeholders, and it may be divided into three relevant time periods.

- The emplacement period. A licence is granted that authorises the transfer and emplacement of waste packages into pre-excavated galleries, rooms, and/or bore-holes. The environmental conditions are monitored continuously and compared with the baseline data. Research and development continues. The regulator performs regular inspections of the underground operations. The long-term safety case is updated regularly by the developer/implementer and reviewed by the regulator. In this phase, new underground galleries may be built, and partial backfilling and/or sealing of galleries and disposal facility areas may also take place.
- The observation period. After all waste packages are emplaced, it might be decided to monitor (parts of) the disposal facility and to keep some accessibility to at least part of the waste while additional performance confirmation takes place.
- The closure period. A licence to close is granted, and complementary backfilling and sealing are performed according to design. Access from the surface to the underground facility is terminated. Surface facilities may be dismantled. All relevant information is preserved in an archive, and society may be involved regularly in oversight of the disposal facility.

The postoperational phase

(31) During this phase, the presence of man is no longer required to directly manage the facility. This is the longest phase, and is divided into two relevant time periods.

- The period of indirect oversight. After closure, safety is assured totally through the intrinsic, built-in safety provisions of the design of the disposal facility. Nevertheless, it is expected that monitoring of the baseline environmental conditions will continue for a period of time as well as regulatory or societal oversight. Archives of technical data and configuration of waste packages and the disposal facility are kept, and the use of warning signs or markers to remind coming generations of its existence may be considered. The relevant international safeguards and controls continue to apply. Inadvertent human intrusion in the disposal facility can be ruled out.
- The period of no oversight. It is not possible to foresee the point at which indirect oversight might terminate; nevertheless, it must still be considered in the design and planning stage as there is no guarantee that the memory of the site will persist into the distant future, and therefore that the oversight will be maintained indefinitely. Eventually, loss of memory and consequently loss of oversight may take place, either progressively or following major unpredictable events such as war or loss of records. Therefore, inadvertent human intrusion in the disposal facility cannot be ruled out during this time period. The intrinsic hazard of the waste decreases with time, but it may continue to pose a significant hazard for a considerable time. Nevertheless, the loss of oversight does not result in a change in the intrinsic protective capability of the disposal facility.

(32) During the operational phase (time of direct oversight), it will be possible to evaluate the protection capability of the disposal facility based on regular updates of the safety case. The safety case provided by the developer of a disposal facility must address the operational and the postoperational phases and, specifically, the distant future when controls and interventions cannot be relied upon. The aim of the developed safety case is to provide evidence of the protective capability of the system that is sufficient to make decisions on its future development or amendment of plans. The safety case shows how the barriers in the disposal facility and associated host rock (the disposal system) work together, and how they fulfil their desired functions over time. It documents the principles and strategies that were followed for developing the knowledge base. It recognises the residual uncertainties in both the long-term processes and potential future events that may affect the performance of the disposal system, and why these have been considered not to reduce protection unduly. Interactions with the various stakeholders (e.g. the local public, outside experts brought in to conduct technical reviews) are acknowledged elements to enhance the quality of the decision-making process at the different phases of development and implementation of the disposal facility.

(33) After closure of the facility (postoperational phase), direct oversight will end and indirect oversight will be performed to provide additional assurance of safety (e.g. environmental monitoring for potential releases to the biosphere and monitoring of relevant performance of the repository). Regulatory oversight may also continue for a period of time. Other administrative bodies may be created to follow up the project and society may also be involved. It is not possible to guarantee that indirect oversight will continue indefinitely into the distant future; however, geological disposal facilities are developed and designed not to rely on the presence of oversight in the distant future but rather to rely on passive safety.

(34) If oversight ceases to exist in the postclosure period, the repository is still a functioning facility and continues to be so. The potential to contain and isolate the radioactive waste is an inherent feature of the radioactive waste repository that continues into the distant future, and responds to the considered evolution of the disposal system under natural processes and events. The multibarrier, multifunction system that is at the basis of the disposal system design must have the potential to constrain releases of radionuclides from the radioactive waste repository. If indirect oversight ceased to exist, there is the possibility that inadvertent human intrusion into the facility might occur. The location of the disposal facility, deep underground, isolated from the environment that humans normally inhabit and in a geological environment with no exploitable resources, together with its technical design, provide protection against inadvertent human intrusion.

3.3.2. Relevant time frames for radiological protection

(35) As stated above, the scope of this report is the description of how protection criteria can be used in the safety assessment, and to establish recommendations on protection issues related to the disposal of long-lived solid radioactive waste. One of the important factors that influence the application of the protection system over

the different phases in the life time of a disposal facility is the level of oversight or 'watchful care' that is present: direct, indirect, and no oversight. The level of oversight directly affects the capability to reduce or avoid some exposures.

(36) During the time of direct oversight, the operator, overseen by the regulator, in interaction with the concerned stakeholders are able to actively manage the protection of workers, the public, and the environment through a set of actions. The transition from this time frame into the time frame of indirect oversight is not abrupt. Thus, parts of the disposal facility are under direct oversight, and at the same time, other parts of the disposal facility are under indirect oversight.

(37) During the time of indirect oversight, there might be some activity of people/staff/operators at the site. Knowledge is maintained, monitoring may continue to occur, and some corrective actions could be made if necessary. As time progresses, the degree of indirect oversight may change, corresponding, for example, to less frequent inspections or the ending of regulatory supervision. The decisions to reduce the level of oversight may be based, to some extent, on the degree of confidence in the behaviour of the facility, and other societal and economic factors. Decisions related to the organisation and evolution of oversight should be discussed with the stakeholders concerned.

(38) In the postoperational period, after the end of active regulatory oversight, maintaining indirect oversight and memory of the facility should become a societal responsibility, possibly discharged through national or local government. One might expect that society will maintain forms of indirect oversight and memory as long as possible. However, there is no guarantee to maintain them in the distant future.

(39) One means to continue oversight after active regulatory oversight has ceased is the preservation of memory or records of the presence of a geological disposal facility. Other measures, such as restrictions on land use, decided by the authorities in interaction with the different stakeholders might also continue to apply. Measures to preserve the memory of a facility might help to reduce the probability of inadvertent human intrusion, and may assist the justification and planning of any deliberate intrusion should this be required in the future. At some point in the distant future, the memory of the presence of the disposal facility may be lost. The choice of location of the geological disposal facility and its technical design will constitute the remaining 'built-in control' against inadvertent intrusion into the facility.

(40) Some national approaches plan emplacement and backfilling strategies that will result in direct oversight of the site lasting for several tens of years after the start of operations. It is not possible to know the criteria that may be used by the people making decisions in the future. The different decisions to be made relating to the evolution of oversight should be discussed with stakeholders.

4. APPLICATION OF THE ICRP SYSTEM OF PROTECTION DURING THE LIFE OF A GEOLOGICAL DISPOSAL FACILITY

(41) The major features of the 2007 Recommendations (ICRP, 2007) relevant to this report are:

- evolving from the previous process-based protection approach using practices and interventions, by moving to a situation-based approach applying the fundamental principles of radiological protection to all controllable exposure situations, which the 2007 ICRP Recommendations characterise as planned, emergency or existing exposure situations.
- maintaining the Commission's three fundamental principles of radiological protection – namely justification, limitation, and optimisation – and clarifying how they apply to sources delivering exposure and to individuals receiving exposure;
- re-enforcing the principle of optimisation of protection, which should be applied in a similar way to all exposure situations, with restrictions on individual doses and risks, namely dose and risk constraints for planned exposure situations, and reference levels for emergency and existing exposure situations.

4.1. Exposure situations

(42) The Recommendations in *Publication 103* organise the system of protection according to three types of exposure situations: planned, existing and emergency situations (ICRP, 2007, Para. 176).

- Planned exposure situations are everyday situations involving the operation of deliberately introduced sources. Planned exposure situations may give rise both to exposures that are reasonably anticipated to occur (normal exposures) and to exposures which may arise following deviations from planned operating procedures as well as exposures to be delivered in the distant future for which there are large uncertainties on their assessment (potential exposures). These potential exposures can be considered at the planning stage.
- Emergency exposure situations are exposure situations resulting from a loss of control of a planned source, or from any unexpected situation (e.g. a malevolent event), which require urgent action to avoid or reduce undesirable exposures.
- Existing exposure situations are situations resulting from sources that already exist when a decision to control them is taken (natural radiation, past activities or after emergencies).

(43) The Commission views the potential exposures to humans and the environment associated with the expected evolution of the geological disposal of long lived solid radioactive as a planned exposure situation. The management of the source is deliberate and clearly planned and there is obligation to provide controls to ensure that during the operation and postoperational phases of a geological disposal facility adoptimized protection is ensured. However, particular circumstances, which may

not be part of the normally expected and planned activities may rise. They are discussed below.

4.2. Fundamental radiological protection principles

(44) The definitions of the three fundamental principles and basic considerations for their application to waste disposal are described as follows.

- The Principle of Justification: 'Any decision that alters the exposure situation should do more good than harm.' Any practice that gives rise to exposure situations needs to be justified. The Commission has previously stated (ICRP, 1997b) that radioactive waste management and disposal operations are an integral part of the practice generating the waste. It is wrong to regard them as a free standing practice that needs its own justification. Therefore, justification of the practice should include the management options of the waste generated, e.g. geological disposal. The justification of a practice should be reviewed over the lifetime of that practice whenever new and important information becomes available: such information may arise for societal, technical and scientific reasons. If the management of waste was not considered in the justification of a practice that is no longer into operation, the Commission recommends to optimize the protection of humans and the environment independently of considering the justification of such practice.
- The Principle of Optimisation of Protection: 'The likelihood of incurring exposure, the number of people exposed, and the magnitude of their individual doses should all be kept as low as reasonably achievable, taking into account economic and societal factors.' As clearly stated in ICRP *Publication 103*, optimisation is of primary importance and its role has been reinforced. This is also the key principle guiding the application of the ICRP system of protection to the disposal of long-lived solid radioactive waste, as discussed in this report (for details see section 4.8). The practical application of this principle includes the use of source-specific dose constraints
- The Principle of Application of Dose Limits: 'The total dose to any individual from regulated sources in planned exposure situations other than medical exposure of patients should not exceed the appropriate limits specified by the Commission.' For radioactive waste management, the general statement of ICRP *Publication 81* (ICRP, 1998, Para. 36) still applies: 'Although the Commission continues to recommend dose limits, it recognises that 'dose limits for public exposure are rarely limiting in practice' (ICRP, 1997b, Para. 36). Furthermore, it considers that '...the application of dose limits to waste disposal has intrinsic difficulties' (ICRP, 1997b, Para. 19) and that control of public exposure through a process of constrained optimisation will obviate the direct use of the public exposure dose limits in the control of radioactive waste disposal' (ICRP, 1997b, Para. 48).'

4.3. Dose and risk concepts

(45) The main and primary use of the effective dose in radiological protection for both occupationally exposed workers and members of the public is for the prospective assessment of dose for planning and optimisation of protection and the retrospective assessment of dose for demonstrating compliance with dose limits, or for comparing with dose constraints or reference levels (ICRP, 2007, Para. 153). In practical radiological protection applications, effective dose is used for the demonstration of compliance with protection standards.

(46) A potential exposure is an exposure that is not expected to be delivered with certainty but that may result from an accident at a source or an event or sequence of events of a probabilistic nature, including equipment failures and operating errors (ICRP, 1993, 1997a). The risk associated with such an event is a function of the probability of an unintended event causing a dose, and the probability of detriment due to that dose. Risk constraints are applied to potential exposures, when applying an aggregated approach combining probability of a dose multiplied by the probability of the resulting health effect. For potential exposures of workers, the Commission continues to recommend a generic risk constraint of 2×10^{-4} year^{-1}. For potential exposures of the public, the Commission continues to recommend a risk constraint of 1×10^{-5} year^{-1}, or to apply the dose constraint in case of adoption of a disaggregated approach for dose and probability of scenarios. Due to the considerable uncertainties surrounding exposures that may arise in the future they are also considered as potential exposures.

(47) For the periods of indirect oversight and no oversight of a geological disposal facility, radiation exposures are treated as potential exposures. The results of estimating risk over long periods of time should be interpreted cautiously, because of the inherent uncertainties in the assumptions. The assessment of post operational radiological impacts through the estimation of risk to a reference person can only provide an indication or illustration of the robustness of the system, rather than predictions of future radiological consequences. For the purpose of optimisation, numerical assessments of risk should be compared with the numerical values of the risk constraint, but it must be recognised that this comparison does not imply strict regulatory compliance with a constraint. For these potential exposures, the Commission also recommends that the value of the risk constraint should be 1×10^{-5} year^{-1}. Strict application of numerical criteria at very long times into the future may be inappropriate.

(48) When considering extremely rare events beyond the expected set of evolution scenarios considered for a facility during the periods of indirect oversight and no oversight, it may be appropriate to estimate the potential radiological impact using stylised or simplified scenarios. The results of those analyses can, if required, be used as indicators of system robustness by comparing them with numerical values of dose or risk. If this approach is adopted, the application of the reference levels defined for existing and/or emergency exposure situations is recommended. Consideration of extremely unlikely events and human intrusion is normally kept separate from

potential exposures from design basis scenarios. The treatment of extremely rare events could vary between sites and between different national approaches.

4.4. Protection in the operational phase

(49) Waste emplacement activities are subject to the same principles of dose limitations and the requirement to optimise below constraints as those in any nuclear facility. Both occupational and public exposures are expected from the transportation, handling, and disposal activities, and thus are planned exposures including potential exposures from deviations from the normal operations. The possibility also exists for incidents due to low-probability/high-consequence-initiating events, some of which may lead to an emergency situation. Operations would be expected to be optimised consistent with the Commission's 2007 Recommendations. The annual dose limit for workers of 20 mSv year^{-1}, averaged over a 5-year period, is applied with the requirement of optimising protection below dose constraints. The recommended dose constraint for the public is 0.3 mSv year^{-1} for each source.

(50) For a typical disposal facility, the safety assessment would suggest that significant releases are unlikely during the emplacement period and the period of time during which a competently sited, operated, and sealed disposal facility is being actively observed and monitored. Therefore, some exposures would be categorised as part of the potential exposure subset of planned exposures, due to accidents, and the rest would be categorised as normal exposures.

4.5. Protection in the postoperational phase

(51) At the end of the period of direct oversight, occupational exposures should be considered in two limited areas of exposure: (1) exposures for any indirect monitoring of the facility and its surroundings during the period of indirect oversight (mainly environmental monitoring); and (2) exposures due to residual radioactivity after decommissioning of the surface facilities. Given the potentially vast time periods involved in the postoperational phase, the possibility of an eventual release of some radioactive substances is inherent in the concept of geological disposal, even if the system operates as intended (i.e. without deviations from procedures in operations, construction, or accidents), and this may lead to exposure of the public. These very-long-term potential releases of radioactive substances and subsequent exposures are assumed to result from a variety of scenarios. While they may be foreseen and perhaps assigned a probability, they are still intrinsically uncertain. Evaluations of these exposures serve the purpose of comparing alternative facility design options, and reaching a regulatory judgement regarding the capability of the system to contain and isolate the waste. Such evaluations are not considered to be predictions, nor are they intended to be used for the protection of specific individuals or populations. Such exposures may, in fact, be projected to occur at such distant times that established concepts such as effective dose and the associated radiological risk have to be used with caution. As stated in *Publication 103*: '(...) dose estimates should not be regarded as measures of health detriment beyond times of around several hundreds

of years into the future. Rather, they represent indicators of the protection afforded by the disposal system' (ICRP, 2007, Para. 265).

(52) Any releases would be expected to take place well beyond the operational period of the facility, so the immediate causes of any release would be beyond the control of the operator. The timing and magnitude of such releases are not predictable except in the broadest sense, using best-available approaches, and hence they are treated as potential exposures. The presence of exposed populations at the point of release, as well as their capability to implement protective and/or corrective actions in the distant future, cannot be assumed with any certainty, should such releases occur.

(53) The process of evaluating the potential exposure from emplaced waste includes understanding the potential mechanisms of radionuclide release from the engineered facility, including modelling transport through the geosphere to the biosphere, and the resultant release into an appropriate environmental compartment that could give rise to exposures to humans and the environment. Depending on the level of knowledge, probabilities may be estimated for these release scenarios. However, at the long time scales considered in geological disposal, evolution of the biosphere and, possibly, the geosphere and the engineered system will increase the uncertainty of these probabilities. Hence, the results of any dose or risk assessments need to be interpreted in a qualitative way at long time scales.

(54) The expected evolution of a geological disposal facility in the distant future should not require active involvement to mitigate the consequences, as this is counter to the principle of avoiding an undue burden on future generations. Therefore, the Commission continues to support its recommendations in *Publication 103* (ICRP, 2007) that either a dose constraint of 0.3 mSv year^{-1} (for the expected evolution scenario) or an annual risk constraint of 1×10^{-5} be used for potential exposures from the emplaced waste.[2] Two approaches may be considered: (1) aggregation of risk by combining doses and probabilities, and comparing the result with the risk constraint; or (2) for each exposure, presenting the dose and its corresponding probability of occurrence separately, and comparison with the dose constraint supplemented by consideration of the probability that the doses would be incurred. As noted in *Publications 81, 82, 101*, and *103* (ICRP, 1998, 1999, 2006, 2007), the Commission considers that although a similar level of protection can be achieved by these two approaches, more information may be obtained to reach risk-informed decisions from separate consideration of the probability of occurrence of a particular situation giving rise to a dose, and the resulting dose. In addition, it should be noted that the disaggregated approach does not require precise quantification of the probability of the occurrence of scenarios, but rather an appreciation of their radiological consequences balanced against the estimated magnitude of their probability.

---

[2] Although regulatory control of a geological disposal facility is not envisaged to continue indefinitely, the disposal of long-lived hazardous waste in a geological disposal facility is a totally different concept to exemption or clearance of waste from regulatory control, and hence the dose criteria for clearance do not apply.

(55) In the distant future, in the case where oversight provisions are no longer operational in case the memory of the presence of the disposal facility is lost, it is possible that people will 'rediscover' the disposal facility. This may be without compromising its integrity (e.g. remote sensing), by observing very small discharges into the biosphere, or it may be by directly breaching the containment, albeit inadvertently, and causing contamination of the environment. Under current guidelines, situations of this type would be treated as existing exposure situations, and it may be postulated that a similar approach will apply in the future.

4.6. Protection in particular circumstances

4.6.1. Natural disruptive events

(56) The disposal facility and its surrounding environment could be impacted or altered by natural events (e.g. earthquake) during the periods of indirect oversight or no oversight. Different scenarios can be envisaged in the future according to current knowledge. For these events, it may be possible to estimate or bound the probability and time frame of occurrence, and the resulting health consequences should be taken into account in reaching risk-informed waste management decisions. These natural events are normally included in the envelope of design-basis scenarios.

(57) Natural disruptive events with very low probability compared with the design basis may occur, and some of these may induce significant disturbances on the disposal facility or the migration of the radioactive substances. Examples of these types of events may include some major landform change due to tectonic events, meteorite impacts, etc. The Commission recommends that a strategy should be developed for addressing natural disruptive events that could result in significant exposure of people and the environment with the involvement of relevant stakeholders. Possible approaches include the establishment of a methodology for excluding low-probability events from consideration in the risk-assessment process, selecting a site with characteristics that minimise the probability of such events, or assessing specific events through stylised assessments.

(58) Previously, the Commission considered all natural events, disruptive or not, within the same framework (ICRP, 1998). Now, the Commission recommends that the two different types of natural events should be considered separately. For natural events that are included in the design-basis evolution, the Commission recommends application of the risk constraint or the dose constraint for planned exposure situations. For severe natural disruptive events not taken into account in the design-basis evolution, application of the risk constraint or the dose constraint for planned exposure situations does not apply. If the events were to occur when there is still (direct or indirect) oversight of the disposal facility, the ensuing situation would be considered by the competent authority and the relevant protection measures would be implemented. If *Publication 103* (ICRP, 2007) is still relevant at the time, it is expected that the reference levels for emergency exposure situations would be applied followed by those for existing exposure situations. If such a disruptive event were to occur when oversight of the disposal facility has ceased, there is no certainty that a competent

authority would be aware of the disturbance or understand the source of exposure. Therefore, it is not possible to consider with certainty the implementation of relevant countermeasures to control the source. If the authorities eventually became aware of the disturbance, action appropriate to the regulatory standards of the time would be expected.

(59) The potential impact of severe disruptive events may be estimated at the design stage using stylised or simplified calculations. An indication of the robustness of the system could be obtained by comparing these results with numerical values of dose, or risk, if required. If this approach was adopted, the appropriate reference levels would be those for an existing exposure situation, or for an emergency exposure situation, depending on the specific scenario. It should be noted that the optimum design of a disposal facility may result in a distribution of doses from severe disruptive events where some could be above the reference level.

(60) For emergency exposure situations, the Commission recommends a reference level in the range of 20–100 mSv, and the development of protection strategies to reduce exposures as low as reasonably achievable below the reference level, taking economic and societal factors into account (ICRP, 2009a).

(61) According to *Publication 103* (ICRP, 2007), long-lasting exposures resulting from natural disruptive events (with or without an emergency phase) should be referred to as 'existing exposure situations', and the recommended reference level for optimising protection strategies ranges between 1 and 20 mSv year^{-1}. In agreement with the Commission's recommendations in *Publication 111* (ICRP, 2009b), a reference level should be selected in the lower part of the band (e.g. in the range of a few mSv per year).

4.6.2. Inadvertent human intrusion

(62) Waste is disposed of in a geological disposal facility for the purposes of containment and isolation (one aspect of which is avoidance of human intrusion). It is necessary to distinguish between deliberate and inadvertent human intrusion into the facility. The former is not discussed further in this report as it is considered outwith the scope of the responsibility of the current generation to protect a deliberate intruder (i.e. a person who is aware of the nature of the facility). The design and siting of the facility have to include features to reduce the possibility of inadvertent human intrusion.

(63) A release resulting from inadvertent human intrusion, such as drilling into the facility, could migrate through the geosphere and biosphere, resulting in exposures that are indirectly related or incidental to the intrusion event. It is also possible that inadvertent human intrusion could bring waste material to the surface, and hence lead to direct exposure of the intruder and nearby populations. This introduces the possibility of elevated exposures and significant doses, which is an inescapable consequence of the decision to isolate and concentrate the waste rather than diluting or dispersing it.

(64) Protection from exposures associated with human intrusion is best accomplished by efforts to reduce the possibility of such events. These may include siting

a disposal facility at great distance from the surface, avoiding assumed valuable resources, incorporating robust design features that make intrusion more difficult, or from existing provisions for indirect oversight (such as restrictions on land use, environmental monitoring programmes, surveillance under safeguards agreements, archived record and site markers). While the actual probability of inadvertent human intrusion at a specific site is largely unknowable as it is based on future human actions, it is assumed that the probability of inadvertent intrusion during the direct and indirect oversight periods is extremely low, and that if it occurred, appropriate countermeasures could be taken to avoid significant impact.

(65) In the distant future, if indirect oversight has ceased, the occurrence of human intrusion cannot be excluded. Therefore, the consequences of one or more plausible stylised intrusion scenarios should be considered by decision makers to evaluate the resilience of the disposal system to potential inadvertent intrusion. Any estimates of the magnitude of intrusion risks are, by necessity, dependent on assumptions that are made about future human behaviour. As no scientific basis exists for predicting the nature or probability of future human actions, the Commission continues to consider it inappropriate to include the probabilities of such events in a quantitative performance assessment that is to be compared with dose or risk constraints (ICRP, 1998). At the planning stage, the results of the stylised or simplified calculations can, if required, be used as indicators of system robustness by comparing them with numerical values of dose. If this approach is taken, the application of the reference levels defined for emergency and/or existing exposure situations is recommended. It should be noted that the optimum design of a disposal facility may result in a distribution of doses from inadvertent human intrusion where some could be above these reference levels. Once an event has happened, there is no certainty that a competent authority would be aware of the disturbance. If the situation is recognised, the competent authority would assess the situation and apply the appropriate protection criteria and countermeasures. If *Publication 103* (ICRP, 2007) was still extant at the time, it is expected that the reference levels for emergency and/or existing exposure situations would be used, as appropriate. In circumstances where doses are estimated to exceed these reference levels, reasonable efforts should be made to reduce the probability of human intrusion or to limit its consequences.

(66) In the case of geological disposal, intrusion means that many of the barriers that were considered in the optimisation of protection for the disposal system have been by-passed. As a future society may be unaware of exposures resulting from such events, any protective actions required should be considered during the development of the disposal facility through siting and design of a geological repository. Furthermore, evaluation of the robustness of the disposal system against human intrusion (see Para. 65) can increase confidence in its safety case.

4.7. Summary of relevant exposure situation according to oversight

(67) Application of the three exposure situations and of dose limits, dose constraints, and reference levels as defined in *Publication 103* (ICRP, 2007) during these time frames is indicated in Table 4.1. Table 4.1 identifies the criteria that ICRP

Table 4.1. Radiological exposure situations as function of disposal facility evolution, and presence and type of oversight.

Disposal facility status	Type of oversight		
	Direct oversight	Indirect oversight	No oversight
Design-basis[*] evolution	Planned (normal and potential) exposure situation[†]	Planned (potential) exposure situation[†,‡]	Planned (potential) exposure situation[†,‡]
Non-design-basis evolution[§]	Emergency exposure situation at the time of exposure, followed by an existing exposure situation	Emergency exposure situation at the time of exposure, followed by an existing exposure situation[¶,**]	Emergency and/or existing exposure situation, once exposure is recognised[¶,**]
Inadvertent human intrusion	Not relevant	Not relevant	Emergency and/or existing exposure situation, once exposure is recognised[¶,**]

[*] The design basis is the envelope of both normal and potential exposures that are used in designing the facility.

[†] In the planning phase: 20 mSv year^{-1} dose limit to workers and dose constraint as specified by the operator; 1 mSv year^{-1} dose limit for public exposures from all sources and 0.3 mSv year^{-1} dose constraint for waste disposal. For potential exposure of the public in case of the application of an aggregated approach, a risk constraint of 1×10^{-5} year^{-1} is recommended.

[‡] No worker dose is foreseen during the period of indirect or no oversight. Releases in the distant future give rise to potential exposure (ICRP, 2007, Para. 265). Comparisons with the dose or risk constraint become increasingly less useful for compliance purposes at times further in the future.

[§] Non-design-basis evolutions include very unlikely or extreme events that could result in significant exposure to humans and the environment.

[¶] If such an event were to occur in the future, the competent authorities of the time would assess whether it had resulted in an emergency exposure situation or in an existing exposure situation, or the equivalent categories of exposure at that time. If *Publication 103* (ICRP, 2007) was still extant, it would be expected that the reference levels for emergency and/or existing exposure situations would be applied, as appropriate. In the period of no oversight, the exposure may not be recognised immediately.

[**] At the planning stage, the potential radiological impact is typically evaluated using stylised or simplified scenarios. The results of those analyses can be used as indicators of system robustness by comparing them with numerical values. In that case, application of the reference levels defined for emergency and/or existing exposure situations is recommended. It should be noted that a fully optimised system may result in a distribution of doses where some are above the reference level (ICRP, 2009a, p. 37).

recommends for comparison during the pre-operational and operational phases. Interventions may be implemented during these periods, if necessary. It also identifies the protection systems that apply during the three main time frames.

(68) The design basis considers a range of incidents, accidents, and natural events, and attempts to ensure that these events are prevented if possible and/or consequences are mitigated.

4.8. Optimisation of protection and Best Available Techniques

(69) The principle of optimisation is defined by the Commission (ICRP, 2006, 2007) as the source-related process to keep the likelihood of incurring exposures (where these are not certain to be received), the number of people exposed, and the magnitude of individual doses as low as reasonably achievable, taking economic and societal factors into account. The general recommendations for the optimisation process are described in *Publication 101* (ICRP, 2006).

(70) The ICRP principle of optimisation of radiological protection when applied to the development and implementation of a geological disposal system has to be understood in the broadest sense as an iterative, systematic, and transparent evaluation of options for enhancing the protective capabilities of the system and for reducing impacts (radiological and others).

(71) Optimisation of protection has to deal with the main aim of disposal systems, i.e. to protect humans and the environment, now and in the future, by containing the radioactive substances in the waste to the largest extent possible and by isolating the waste from man, the environment and the biosphere. Optimisation of protection has to deal with the protection of workers, the public and the environment during the time of operation, as well as with the protection of future generations including possible periods of no oversight. In the long term and particularly, in the latter period, safety has to be ensured by the legacy of a passively functioning disposal system.

(72) The stepwise decision process for geological disposal system development and implementation constitutes the framework for the optimisation process. As a central component, optimisation and the application of Best Available Techniques have to cover all elements of the disposal system in an integrative approach [i.e. site (including host rock formation), facility design, waste package design, waste characteristics] as well as all relevant time periods.

(73) Optimisation of protection is the responsibility of the developer, and involves liaison with safety and environmental protection authorities, local communities, and other stakeholders; multiple decisions have to be taken. Therefore, it is not possible to define, *a priori*, the path for a sound optimisation process for a geological disposal system, or the success criteria for the end result of an optimisation process.

(74) Socio-economic factors (including policy decisions and societal acceptance issues) can bound the optimisation process to various extents, such as by limiting the available options (e.g. siting) and/or by defining additional conditions (e.g. retrievability). It is important that these considerations are identified in a manner transparent to all involved stakeholders, and that their safety implications are generally and broadly understood (OECD/NEA, 2011).

(75) Although optimisation is a continuous effort, milestones have to be defined in the stepwise process, where all involved stakeholders can judge the result of the optimisation process and indicate ways to improve various elements of the system.

(76) The process of optimisation is considerably different for the pre-operational, operational, and postoperational phases. During the operational phase, the general recommendations for any large nuclear facility apply. Experience gained during the

operational phase can be factored into immediate or near-term improvements, reducing the exposure to both workers and the public from the emplacement work.

(77) Nearly all aspects of optimisation for the postoperational phase must occur prior to waste emplacement, largely in the siting and design phase, with the plans to close the facility being part of the design phase. Some further optimisation of the protection that will be provided during the postoperational phase is still possible during the operational phase; for example, new materials or techniques may become available. Experience gained during the closure of parts of the facility (e.g. sealing of disposal rooms) can lead to improvements in planning for the closure of the overall facility.

(78) During the postoperational phase, there is no active operation of the disposal system. The waste is emplaced and the protection of humans and the environment is mainly based on the passive isolation and containment capabilities of the disposal system. Hence, decisions on optimisation in the postoperational phase can only relate to provisions of indirect oversight of the closed disposal system.

(79) Geological disposal facilities are sited, designed, and implemented to provide for robust long-term isolation and containment, resulting in potential impacts on humans and the environment only in the very distant future. Consequently, as explained earlier, the assessment of postclosure radiological impacts through the estimation of effective dose or risk to a reference person, given the increasing uncertainties with time and the cautious assumptions to be made, can only provide an indication or illustration of the robustness of the system, rather than predictions of future radiological consequences. Thus, when considering the distant future, dose and risk values lose their intrinsic meaning and only retain a value as relative comparators of potential radiological impact.

(80) The elements guiding or directing the optimisation process should be those that directly or indirectly determine the quality of the components of the facility as built, operated, and closed, where quality refers to the capacity of the components to fulfil the safety functions of containment and isolation in a robust manner. The assessment and judgement of the quality of system components essentially includes the site characteristics, elements of Best Available Technique, as well as the concepts of good practice, sound engineering, and managerial principles. These elements complement and support radiological optimisation when potential impacts in the distant future have to be dealt with.

(81) Judgement of the quality of the system design developed or implemented has to be made, and reviewed critically when needed, in a well-structured and transparent process, with the involvement of all relevant stakeholders. At the heart of this process is the interaction, transparent for all other stakeholders, between the developer and the safety authorities.

(82) When dealing with safety in the more distant future, optimisation can be complemented and supported by applying the concept of Best Available Technique on the various levels of the disposal system, through:

- the methodologies for identifying and selecting the methodological and scientific programme of site characterisation in order to assess its containment and isolation capacities now and in the distant future;

47

- the development of the system design, including the choices of materials and technologies, and the way they will contribute, individually and together, to the main aim of containment and isolation, taking due account of the characteristics of the site;
- the integration of waste, site, and design characteristics within one disposal system and the iterative assessment of the containment and isolation capacities of the system as a whole; and
- the use of sound managerial and engineering methods and practices during system construction, operation, and closure, within an integrated management system.

(83) Optimisation on the basis of radiological criteria (effective dose and risk) is an important part of the optimisation of the design and implementation process of the disposal system at specific periods and for specific aspects of the disposal system (e.g. when operational safety is assessed during the design development steps, and during preparation and implementation of operational procedures and activities).

(84) The way in which the various elements of a disposal system can be optimised in an integrative manner during system development varies to a large extent. First of all, stepwise optimisation decisions mainly have to be taken in chronological order (e.g. the decisions on the choice of a host rock and on one or a limited number of sites are often prior to decisions on a detailed design). For the selection of a site, a balance has to be struck between technical criteria related to the safety of a disposal system (long-term stability, barrier for radionuclide migration, absence or presence of natural resources in the vicinity), and local or supralocal economic and societal factors. Favourable sites can, in a first step, be identified on the basis of broadly defined 'required qualities', taking due account of the containment and isolation function(s) of the natural barriers and the natural environment in the disposal system.

(85) If several suitable sites can be identified and evaluated, the decision in favour of one specific host rock or site will always be a multifactor decision, based on both qualitative and quantitative judgements. Radiological criteria (e.g. calculated effective dose or risk) are often of limited value for this multifactor decision due to: (1) the increasing uncertainties for longer assessment time scales, and (2) the observation that calculated radiological design-basis impacts are often so low that they do not constitute a discriminating factor for the choice of a site.

(86) Assessment of the robustness of the disposal system can contribute to system optimisation because it provides insight, quantitative or qualitative, into the performance of the disposal system and its components, and into the relative contributions of the various components to the overall system. Therefore, the value of such an assessment for the optimisation process is mainly through the insights it provides on the relative contributions of the various components to the overall system objectives of containment and isolation, and how these contributions can be affected by disturbing events and processes or by remaining uncertainties. The uncertain nature of calculated effective dose and risk that is estimated to arise in the very distant future reduces their usefulness for the optimisation process.

4.9. Technical and management principles and requirements

(87) The general implementation of the Commission's recommendations on the disposal of radioactive waste requires that organisational and managerial structures and processes are put into place, and that technical principles are applied. Organisational structures and processes can differ between countries, but should be based on the principles laid down by the International Atomic Energy Agency in its fundamental safety principles and safety standards on management systems (IAEA, 2006, 2009).

(88) The Commission recommends that management principles and requirements should be applied to the disposal system development and implementation process to enhance confidence that the protection of humans and the environment will be ensured for as long as needed. This requires the implementation of a management system that integrates safety, health, environmental, security, quality, and economic elements, with safety being the fundamental principle upon which the management system is based (OECD/NEA, 2007, 2010).

(89) Management systems play an important role to:

- improve, in a systematic manner, the safety performance of an organisation through the planning, control, and supervision of its safety-related activities;
- foster and support a strong safety culture through the development and reinforcement of good attitudes and behaviour in individuals and teams for all safety-related tasks; and
- maintain and further develop knowledge, competences, and skills for the disposal of radioactive waste, as an essential element to ensure high levels of safety; this should be based on a combination of scientific research and technological development, insights gained from successive safety cases, learning through operational experience, and technical cooperation between all actors. Independent reviews, transparency and accessibility of information, and openness to stakeholder participation are also important contributors for ensuring high levels of safety.

(90) A key technical principle for developing disposal systems and assessing their safety is the concept of defence in depth, which provides for successive passive safety measures, enhancing the confidence that the disposal system is robust and has an adequate margin of safety. The defence in depth concept as applied to disposal systems imposes that safety is provided by means of the various components of the system contributing to fulfilling the main safety functions in different ways over different time scales. The performance of the various components contributing to fulfilling the main safety functions has to be achieved by diverse physical and chemical processes, such that the overall performance of the system will not be unduly dependent on a single component or function. The main safety objective of the siting (selecting the natural barrier system and its environment) and designing (developing the manmade barrier system, taking due account of the site characteristics) of a disposal system is to ensure that postclosure safety will be provided by means of multiple safety functions, and that even if a component or safety feature does not perform fully as expected, a sufficient margin of safety will remain.

5. ENDPOINT CONSIDERATIONS

5.1. The representative person

(91) The Commission considers that its recommendations on the estimation of exposures in *Publication 101* (ICRP, 2006) apply as general guidance. The Commission recommends that for planned exposure situations, exposures should, in general, be assessed on the basis of the annual dose to the representative person.

(92) During the operational phase, management of exposures to workers and the public should be the same as for any other facility. During the postoperational phase, due to the long time scales under consideration, the habits and characteristics of the representative person, as well as the characteristics of the host environment, are the product of conjecture. In that case, any such representative person has to be hypothetical and stylised. The habits and characteristics assumed for the individual in the distant future should be chosen on the basis of reasonably conservative and plausible assumptions, considering site- or region-specific information as well as biological and physiological determinants of human life. Moreover, in many cases, different scenarios, each associated with different representative persons, may be considered for the distant future and have different probabilities of occurrence, although establishing discrete probabilities may be problematic. Thus, the scenario leading to the highest dose may not be linked to the highest risk. It is therefore important for decision makers to have a clear presentation of the different scenarios, and either their associated probabilities of occurrence or an appreciation of their corresponding probabilities.

(93) As stated in *Publication 101* (ICRP, 2006), for the purpose of protection of the public, the representative person corresponds to an individual receiving a dose that is representative of the more highly exposed individuals in the population. Therefore, it should be assumed that the representative person is located at the time and place of the maximum concentration of radionuclides in the accessible biosphere, with due regard to the assumed climatic conditions for that evolution scenario (e.g. considerations of ice coverage). This is an assumption as humans may no longer inhabit these areas in the distant future.

(94) A representative person cannot be defined independently of the assumed biosphere. Major changes may occur in the biosphere in the long term due to the action of natural forces in a similar manner to those occurring in the past. Human actions may also affect the biosphere, but one can only speculate about human behaviour in the long term. In the definition of the scenarios, consideration of biosphere changes should be limited to those due to natural forces. A representative person and biosphere should be defined using either a site-specific approach based on site- or region-specific information, or a stylised approach based on more general habits and conditions; the use of stylised approaches will become more important for longer time scales.

(95) The Commission recommends (ICRP, 2006) the use of three age categories for estimation of annual dose to the representative person for comparison with annual dose or risk criteria (note that the annual dose from the intake of a radionuclide

already includes a component relating to the fact that the radionuclide will deliver a dose in successive years, the length of time being determined by the biological half-life of the radionuclide in the body). These categories are 0–5 years (infant), 6–15 years (child), and 16–70 years (adult). In the case of geological disposal, any exposures are expected to occur in the distant future, and to be associated with levels of radionuclides in the environment that change slowly over the time scale of a human life time. Given the inherent uncertainties in calculations extending to the distant future, the dose or risk to an adult representative person will adequately represent the exposure of a person representative of the more highly exposed individuals in the population.

5.2. Protection of the environment

(96) Illustration that the environment is or will be protected against the harmful effects of releases from facilities is an increasing requirement in national legislation, and in relation to many human activities including the management of long-lived waste. ICRP has responded to this need, as well as to a number of other requirements of an ethical nature (ICRP, 2003), by addressing environmental protection directly and specifically in *Publication 103* (ICRP, 2007), and by offering a methodology to address this issue, as outlined in *Publication 108* (ICRP, 2008) and supplemented in *Publication 114* (ICRP, 2009c).

(97) The ICRP approach considers the protection of the environment (not the presence of contamination or other factors that may affect the environment as a resource) by virtue of the aim of 'preventing or reducing the frequency of deleterious effects on fauna and flora to a level where they would have a negligible impact on the maintenance of biological diversity, the conservation of species, or the health status of natural habitats, communities and ecosystems' (ICRP, 2007, Para. 30). The full evaluation of environmental impact would normally be assessed through the environmental impact assessment process and in the environmental impact statement, where effects will be considered within a broader context including factors such as, *inter alia*, visual impact, chemotoxic impact, noise, land use, and impact on amenities.

(98) The default tool for protection and protective actions should be the set of Reference Animals and Plants that has been described by ICRP and for which the relevant data sets have been derived (ICRP, 2008, 2009c). This set was deliberately chosen because its components are considered to be typical biotic types of the major environmental domains of land, sea, and fresh water. As stated earlier in this report, over the long time frames that are considered in waste disposal, the biosphere is likely to change, and may even change substantially. Such changes may entail biosphere evolution with time that is either natural or is enhanced or perturbed through human action. Contributing factors may be climate change, including glaciation cycles, and land uplift or depression. Thus, use of the Reference Animals and Plants should provide at least one point of reference for considering, if necessary, the likely dose and effect in any existing or altered environment in the future.

(99) Thus, the use of Reference Animals and Plants offers, on the one hand, a challenge for waste management which is at least similar to the challenges of demonstrating compliance with dose/risk standards for humans, but, on the other hand, also offers an additional line of argument and reasoning in building a safety case, using endpoints that are different from, but complementary to, protection of human health. Consideration of environmental protection, where appropriate, will thus broaden the basis for risk-informed decision making, and address issues that may have differing levels of importance for different stakeholders.

(100) As stated earlier in this report, over the long time frames that are considered in waste disposal, the biosphere is likely to change, and even change substantially. Such changes entail biosphere evolution with time that is either natural or enhanced or perturbed through human action. Contributing factors may be climate change, including glaciation cycles, and land uplift or depression. Current knowledge of different biospheres and the assessment of impacts on Reference Animals and Plants in those biospheres may aid understanding of potential biosphere changes, and therefore inform decisions related to environmental protection.

6. CONCLUSION

(101) This report explains how the 2007 ICRP system of radiological protection described in *Publication 103* (ICRP, 2007) can be applied in the context of the geological disposal of long-lived solid radioactive waste. It describes the different stages in the life time of a geological disposal facility, and addresses application of the radiological protection principles for each stage depending on the various exposure situations that may be encountered. In particular, the crucial factor that influences application of the protection system over the different phases in the life time of a disposal facility is the level of oversight that is present.

(102) Although many considerations in this report are also relevant to near-surface disposal, it does not supersede previous ICRP recommendations (ICRP, 1998) for the radiological protection of occupationally exposed workers, members of the public, and the environment in the case of near-surface disposal facilities or other disposal options.

REFERENCES

IAEA, 1997. Joint Convention on the Safety of Spent Fuel Management and on the Safety of Radioactive Waste Management. INFCIRC/546. International Atomic Energy Agency, Vienna, Austria.

IAEA, 2006. Fundamental Safety Principles. Safety Standards Series No. SF-1, International Atomic Energy Agency, Vienna, Austria.

IAEA, 2009. The Management System for the Disposal of Radioactive Waste. Safety Standards, Safety Guide No. GS-G-3.4, Vienna, Austria.

ICRP, 1966. Principles of Environmental Monitoring Related to the Handling of Radioactive Materials. ICRP Publication 7. Pergamon Press, Oxford, UK.

ICRP, 1977. Recommendations of the International Commission on Radiological Protection. ICRP Publication 26. Ann. ICRP 1(3).

ICRP, 1985. Principles for the disposal of solid radioactive waste. ICRP Publication 46. Ann. ICRP 15(4).

ICRP, 1991. 1990 Recommendations of the International Commission on Radiological Protection. ICRP Publication 60. Ann. ICRP 21(1–3).

ICRP, 1993. Protection from potential exposure: a conceptual framework. ICRP Publication 64. Ann. ICRP 23(1).

ICRP, 1997a. Protection from potential exposures: application to selected sources. ICRP Publication 76. Ann. ICRP 27(2).

ICRP, 1997b. Radiological protection policy for the disposal of radioactive waste. ICRP Publication 77. Ann. ICRP 27(Suppl.).

ICRP, 1998. Protection recommendations as applied to the disposal of long-lived solid radioactive waste. ICRP Publication 81. Ann. ICRP 28(4).

ICRP, 1999. Protection of the public in situations of prolonged exposure. ICRP Publication 82. Ann. ICRP 29(1/2).

ICRP, 2003. A framework for assessing the impact of ionising on non-human species. ICRP Publication 91. Ann. ICRP 33(3).

ICRP, 2006. Assessing dose of the representative person for the purpose of radiation protection of the public. ICRP Publication 101. Ann. ICRP 36(3).

ICRP, 2007. The 2007 Recommendations of the International Commission on Radiological Protection. ICRP Publication 103. Ann. ICRP 37(2–4).

ICRP, 2008. Environmental protection: the concept and use of reference animals and plants. ICRP Publication 108. Ann. ICRP 38(4–6).

ICRP, 2009a. Application of the Commission's recommendations for the protection of people in emergency exposure situations. ICRP Publication 109. Ann. ICRP 39(1).

ICRP, 2009b. Application of the Commission's recommendations to the protection of people living in long-term contaminated areas after a nuclear accident or a radiation emergency. ICRP Publication 111. Ann. ICRP 39(3).

ICRP, 2009c. Environmental protection: transfer parameters for reference animals and plants. ICRP Publication 114. Ann. ICRP 39(6).

OECD/NEA, 2007. Regulating the Long-term Safety of Geological Disposal – Towards a Common Understanding of the Main Objectives and Bases of Safety Criteria. OECD/NEA, Paris, France.

OECD/NEA, 2008. Moving Forward with Geological Disposal of Radioactive Waste – a Collective Statement by the NEA Radioactive Waste Management Committee. OECD/NEA, Paris, France.

OECD/NEA, 2010. Main Findings in the International Workshop 'Towards Transparent, Proportionate and Deliverable Regulation for Geological Disposal', 20–22 January 2009, Tokyo, Japan. OECD/NEA, Paris, France.

OECD/NEA, 2011. Reversibility and Retrievability (R&R) for the Deep Disposal of High-Level Radioactive Waste and Spent Fuel; Final Report of the NEA R&R Project (2007–2011). OECD/NEA, Paris, France.